THE LITTLE BOOK OF PROJECT MANAGEMENT WISDOM

A Quick Reference for Technology Projects

Robert Forsyth

To my beautiful wife, Nancy and my amazing son Harrison.

CONTENTS

PREFACE

E mployed in the Information Technology field over the past four decades, I've worked with many customers and colleagues on a wide range of projects and programs. Although I spent much time with senior managers, whenever I could, I was down in the "trench" with the operations teams and implementation specialists, doing the most challenging work.

Very little of what I learned was from project management textbooks. Those books covered a lot of theory but avoided talking about many of the real challenges faced by project managers. While the textbooks were useful, my best education came from working with real people on the ground and experiencing things firsthand. After some reflection I decided, through this book, to share some important lessons I learned.

Whether you are a new or seasoned project manager, I hope this book will help you.

WHY YOU SHOULD READ THIS BOOK

I 've managed hundreds of projects through the years. Through success and failures, I learned a lot from my experiences. Some projects were dead-easy and others were an uphill struggle. Why did some projects almost manage themselves, while others had frustrating barriers to overcome and had to be pushed, prodded and cajoled every step of the way to completion?

The answer is that projects are a mix of controllable and less-controllable parts. Like poker hands dealt in a card game, all projects are different. The mix of cards in your hand defines how you approach your objective. The project manager's job is to use all the skills and resources at their disposal to achieve the best outcome. Because we're not born with all the knowledge we need to play the game, it helps to turn to things like this book for guidance.

By reading this book, you'll gain some of the insights that I've acquired through experience, successes, making mistakes, and learning from my clients and mentors. You may already know some of the things I've shared here, but even if you only find one or two helpful things in this book, I'm confident your time will be well-spent.

Designed to be quick and easy to digest, each chapter covers things you'll probably encounter during the life cycle of your projects. I've tried to cover topics with minimal fluff because folks involved with projects are busy and like to get to the point quickly. This book should take no more than a few hours to read and you don't need to read the chapters in any specific order. Hop around if you wish.

Project managers aren't the only ones who will benefit from this book. After reading it, many frustrated project managers will want to discretely place a copy on the desk of their favourite project sponsor. Thus, I've written this guide to help not only project managers, but also executive sponsors, sales folks and other stakeholders who get involved in project development and execution. This book can benefit anyone in a customer-oriented job or activity.

A MESSAGE FOR PROJECT SPONSORS

Many project sponsors are amazing. They're inclusive in their planning and get project managers involved before starting projects. They are engaged from the top-down and have realistic expectations. If you're a sponsor who follows this approach, I salute you. Unfortunately, not all sponsors are like this, so in this chapter, I've included things that could help those "less-than-perfect" sponsors.

The Ivory Tower

Despite many larger clients having in-house PM's on staff, projects are often conceived high up the "ivory tower" without involving project managers and other stakeholders. Senior management, internal middle managers, and vendor salespeople frequently cook-up new projects in isolation. When key stakeholders don't get a chance to participate in the concept phase it creates all sorts of challenges once things get underway. Using examples from real-life projects, the following graphic shows the typical reactions when executives march ahead without inputs from the troops.

Senior VP
"I want Project X rolled out in 6 months, get me a plan!"

Vendor	PMO	Business	IT Operations	Executor	End User
"Yeah, I'm all for it, we can do this, please send me the PO !"	"Arrg ! We need 4 weeks for planning and securing resources, and that target date is unrealistic"	"What ? We have only 6 maintenance windows available in that timeframe, and a 1 month change freeze starts in month 4"	"Hello ! We have no space, no power and limited network capacity"	"Um, we need to upgrade the old system so this will work. An 8 hour outage will be needed , you okay with that ?"	"Wait a second, Solution X doesn't even meet our requirements !"

Let's look at how to avoid this situation.

Things To Consider At Initiation

Projects are often the enabler to achieve new business objectives. Frequently, business objectives are time-sensitive, so there's a rush to get projects started.

Technical Considerations

Technology projects either involve building a brand new technology platform or upgrading an existing one. Virtually all IT projects somehow connect to an existing infrastructure layer, so some form of integration will be required. Integration can be a pesky task because the connection between old and new things always creates risks and challenges.

Business Considerations

The business often rules the schedule. Even with seamless integration, some form of business disruption, like a reboot, could be needed to execute a new solution. Unfortunately, even small disruptions like reboots can be costly to a business. Despite all the urgency in the world, the business unit usually dictates the schedule.

How Long will it take ?
When asked by sponsors how long a technology project will take, I always reply with "it depends." It depends on how we plan to execute the technology-related tasks across the various systems within the "business" constraints. Once the pieces of the puzzle come together, you'll have enough data to develop the schedule.

Get A Project Manager Involved Early

Stakeholders are anyone impacted by or involved in a project. Early participation from stakeholders is vital, so realistic object-ives can be established, and unrealistic targets avoided. If you have one, I suggest involving your in-house PMO or PM when con-ceptualizing a new project. These valuable resources will help you work through the project initiation phase in a structured way and bring the appropriate stakeholders to the table.

Does This Make Sense?

Ask yourself if your next big project makes sense. Does it intro-duce more risk than it solves? Some times replacing a bad spark plug wire is better than replacing the whole engine. Solving a spe-cific problem is often better for your organization than an over-arching project. Share your rationale with those on the ground for "buy-in" and get their full support and constructive input.

Everything Works Perfectly In Powerpoint!

Did you ever notice how solutions rarely roll-out as easy as they looked during the sales presentation? That's because everything works perfectly in PowerPoint! As the next step in the procurement stage, ask for a live demonstration, a proof of concept, or a conference call with the vendor's engineering department before embarking on significant technology investments.

Who Are The Trusted Advisors?

Trusted advisors are those people who give unbiased, candid and realistic assessments of project risks and benefits. The keyword here is "unbiased." Trusted advisors never sugar coat things. They are often people working in the trench with the knowledge and experience you need to help you make informed decisions.

Vendors love to be the client's trusted advisors because this places them in a position of influence. Sometimes a vendor-based trusted advisor provides enormous value because it provides you with a "free" consultant. The danger is they work for someone who wants you to buy their solution, so are they unbiased? Using a vendor based advisor limits you to the constraints of their technology. Why would they point you to a competitor's solution that could work better?

You might look for a trusted advisor inside your organization, or another alternative is to find a vendor-neutral consultant and retain their services. Securing a neutral consultant allows for an unbiased look at multiple solutions and finding the best one. Regardless of your approach, find someone you can trust and listen to them before cooking a project. By doing this one thing, you could save your organization risk, pain and money.

Listen To Your Troops On The Ground

In some firms, the troops view executives as aloof and out-of-touch with reality. They perceive that executives work in their luxurious corner offices, only interested in interacting with other executives. Executives typically interface with the ground troops through their management team, who then disseminate the state of operations back up the chain. This flow of "internal intelligence" is understandable because we know executives are very busy and can't do it all. How many times have you, as an executive or senior manager, visited your troops in the field?

Why not go downstairs and have a coffee with your server admin, storage admin, facilities guy or project manager? Ask them for their honest opinion and listen to their feedback and ideas. They know more about your environment's realities than your managers could ever understand, and they're the ones personally impacted by the decisions you make.

In the least, you'll strengthen your relationship with your team, and you might come back with some exciting new insights or ideas. Avoid making significant commitments that impact your troops without making the trip downstairs and listening to them first.

Make Goals Clear

In May 1961, President Kennedy delivered a dramatic speech to the US Congress; He declared a goal for the United States to land a man on the moon and return him safely by the end of the decade. Despite being the most ambitious technical project in history, it had a crystal clear objective (Land a man on the moon and return him safely) accompanied by a precise deadline (by the end of the 1960s). His project goal was clear and easily understood by anyone from the ground-up. Similarly, your objectives should require as little explanation as possible.

Be Realistic

Defining specific objectives and dates is essential, but are they realistic for your organization? What are the expected results? Do you have the funding, experience and resources to achieve the goal? Imposing unattainable schedules or objectives places a strain on your organization, and will quickly have your troops questioning your judgment.

Don't hold your team accountable to a deadline before examining the project tasks. It's not realistic to ask for a solid project plan or establish end-dates before the project planning starts. During initiation, the tentative date is okay as a target, but give your project team a chance to conduct an assessment and have them suggest a realistic deadline before committing everyone to a schedule they can't possibly achieve. Your people are most likely willing to tackle any ambitious project when it's realistic.

Consider Lead Times And Contracts

When you decide on a project start date, consider materials and resources usually start in motion after the contract is signed and the PO is cut, not before. Your estimated project start and completion dates should consider these factors. Until liability and scope are correctly defined, avoid allowing contractors to commence work before contracts are signed. Starting work without a contract is risky should it touch your business operations or data. For example, if there was a data breach/loss incident caused by someone working without a valid contract, then who's liable? Would your business insurance cover it?

You Get What You Pay For

Placing your technology in the hands of the cheapest bidder could be putting your company at risk. In most cases, you get what you pay for. There's usually a compromise hidden in low-cost products or services, as suggested in the table below:

Factor	By picking the cheapest solution ...	Risk / Cost of Compromise
Performance	The solution is not sized appropriately for your environment	• Performance issues may be discovered after solution is put into production • Business impact $ / outage • End user dissatisfaction • Replacement solution or rework may be needed to remedy poor performance
Features	The solution is missing required features or licenses	• Missing features / licenses may create extra operational costs if needed • Rework may be required
Capacity	The solution is undersized to meet existing or expected growth	• Capacity expansion may be required earlier than budgeted • Extra maintenance windows will be required to upgrade in future
Service Deliverables	Required inclusive or follow on services missing.	• Additional costs above budget • Project delays
Resource Experience	Junior resources will be provided vs. senior resources	• Resources could be learning on your project • Project may take longer due to inexperience • Errors causing business impact
Resource Location	Offshore resources will be provided vs. onsite	• Time zone challenges • Security issues • Communications • May require more client staff

Beware Of Glowing Reports

After the project has started, continue to encourage open communications. No one likes to upset the folks in the Ivory Tower. As an executive, you're likely isolated from what happens out there. Things never go perfectly. Some organizations frown on giving executives open and honest reporting – until it's too late. Be cautious of glowing reports on progress.

Expect Maximum Effort, Accept Challenges

You should expect competent, maximum effort from your project delivery team, but accept that unpredictable things can happen. Things can go wrong in IT, even with the best-laid plans. It won't always be rosy; there may be challenges and setbacks. Problem-handling is a vital part of managing the project and is discussed in detail later in this book

I have this advice for executives; when a problem happens, remain calm! Avoid emotional or punitive actions towards those who are on the support and delivery teams. The team on the ground will have enough pressure without you adding to their stress levels. As the team works through a problem, it's crucial to avoid making threats and giving ultimatums. Allow the team enough runway to do their work and provide any support you can. Expect periodic reports on the remediation progress.

Once, my team ran into a severe issue while running a large enterprise client project. Thankfully the sponsor was a senior storage manager who was a role model for the above approach. We all knew we had encountered a wicked problem, but he remained supportive and level headed during the entire incident. He never showed anger or made threats. His maturity allowed us to focus on working through the problem and not being distracted by doing "damage control." He treated us all like professionals, and we were able to get the issue resolved quickly.

Things that go wrong are also great learning opportunities. You should expect a Root Cause Analysis (RCA) report and team review for significant problems.

Sludge In Your Processes

The planet has gone process crazy. Consultants make a lot of money selling you process ideas and new methodologies. Processes are lovely when needed, but they're like sludge in your car engine; too much of a build-up impacts performance. Despite being created with good intentions, they're often thoughtlessly tacked onto existing procedures to solve quality problems or improve the customer experience. Unfortunately, the poor devils forced to follow the modified processes often suffer painfully. Cumbersome processes add additional workload to your staff and reduce your agility as a company. Was the process executed to fix a problem or cover up a challenge? Are the benefits and costs being measured?

Before introducing a new process, consult with the folks on the ground and get their inputs. There's usually a better way to solve the issue than adding a pile of further steps to someone's already hectic workload. The best process is one with a clear benefit and streamlines workflows.

Your Biggest Assets

You should know this already. Your biggest asset is your people and their wisdom, experience and knowledge. Senior people are leaving organizations as they reach retirement age, taking their priceless wisdom and experience away. Companies often rave how much they value their databases, their security, their systems, but often ignore their most significant asset:

Intellectual Capital

Intellectual capital isn't just knowledge; it's also experience. We can store bits of knowledge in a database, but it's impossible to save "experience." Would you erase years of valuable data from your database archives? Of course not, then why would you allow your most precious knowledge assets to pick up and leave your organization? As the next generation nears retirement, companies should consider retaining their experience through innovative methods such as part-time employment and contracts. However, despite retention tactics, you'll still lose people for other reasons and succession planning for crucial resources is essential.

Trust And Empowerment

When you start taking away empowerment from people, making their jobs robotic, restrictive or transactional, you take their motivation away. An excellent way to inspire your organization is to show employees they're trusted. Allow them to be creative and empower them to make decisions. You don't need to hand them the keys to everything, but your employees are not dummies; they can make right judgment calls on most things. How many organizations have useless layers of management put in place to manage things employees could do themselves?

Outsourcing

I wanted to share a few things I've observed in projects where companies have outsourced IT operations to third-party organizations. There are certainly positive aspects to the outsource approach, but you should also consider the following challenges:

Control
You risk losing control of some of the environment depending on what outsourcing model adopted. Consider what makes good business sense for your organization (SaaS, IaaS, PaaS or complete operations/staff outsourcing) and what controls you want to maintain.

Security
Where will your data be hosted? Will it be physically and electronically secure? How will backup and data recovery be managed?

Complexity
Because of additional management layers involved with outsourced IT, expect to multiply the channels of communications required to get anything done. Instead of a client-vendor project,

you now have a client-provider-vendor project. The formula to calculate the number of channels of communications is N(N-1)/2, where N equals the number of project team members. In connection with this, you will probably need <u>additional layers of change controls</u>.

Nickle and Dime Syndrome

Outsource providers are very transactional when services are delivered. Costs can escalate very quickly ("nickel and dime" scenario). You may start paying for things that you took for granted before outsourcing.

Difficult to Roll Back

In the future, should you want to revert operations to "in-house" IT from an outsource service provider, <u>it may be difficult</u>. Once an outsourcing service provider is in control of your environment, it could be technically challenging and costly to migrate everything back to your organization. Similarly, if you are unhappy with your outsource provider, moving from one to another, could also present challenges and high costs.

Loss of Knowledge / Impact on Staff

Consider the impact on your staff and the loss of intellectual capital. Outsourcing is a sensitive HR challenge. Resentment by those affected and resistance to sharing information could be a possible outcome.

Key Points

☐ Involve project managers/technical experts early.

☐ Knowledge is required before a plan can take shape.

☐ Does the project make sense?

☐ Find an <u>unbiased</u> "trusted advisor."

☐ Visit your troops and listen to their inputs.

☐ Make goals clear, easy to digest and realistic.

☐ Solutions are always simple and work perfectly in PowerPoint.

☐ The lowest price comes with compromises.

☐ Consider lead times/contracts when scheduling.

☐ Remain calm when problems occur.

☐ Be skeptical of glowing reports.

☐ Poorly designed processes hinder productivity.

☐ Your biggest asset is people.

☐ Empower/trust your employees.

☐ Outsourcing could have land-mines.

FOR THE VENDOR SALES FOLKS

A s mentioned in the previous chapter, many IT projects are sold and delivered through external IT vendors. Their focus is to deliver solutions and develop trusted long-term relationships with their clients. In my experience, most vendor sale teams have very high standards and genuinely want to do a good job, but sometimes I found things got a little rough when we had to deliver what was sold. This chapter contains thoughts I'd like to share with the sales folks to help things go smoother.

Project Discovery

Many projects have technical complexities and constraints invisible at the proposal stage without "peeling-back the onion." To reveal an accurate picture, "discovery" work is required, which can be costly because it requires time and resources to accomplish correctly. When projects are sold without vital envronmental data and inputs from stakeholders on the ground, it creates chaos once executed.

Project discovery is essential because it identifies the terrain and obstacles ahead of the planning. Who should fund discovery? Clients don't like to pay for it upfront, and vendors don't want to fund it before the PO is issued. As a result, many times, this critical step is skipped over. Skirting discovery creates false assumptions and timeline expectations.

Sooner or later, some form of discovery will need to take place. When that happens, be prepared to change the project schedule after the inside of the onion has been revealed.

Discovery is ideally started in the initiation phase but flows into the planning phase if the environment is large or complex. Take a look inside the "onion" before planning because it will provide the inputs you need for success. Setting up timelines without proper data places the project at risk of failing to hit objectives and meet expectations.

Get The Right People Involved Up Front

Getting the right people involved in the proposal stage places the project on a solid footing when planning is ready to start.

Three things you must do at the concept stage

1. Engage a project manager in the sales cycle.
2. Engage SMEs from your team and the client to review the viability of the project.
3. Allow the PM's and SME's to participate in sales presentations and proposal meetings, and give them a comfortable forum to share their candid thoughts.

Milestones And Timelines

In the concept/initiation stage, project sponsors always want milestone dates and a project timeline. Please tread very carefully when sharing schedules and milestone dates! Once the sponsor hears any kind of date from you, be assured their ears will perk up, and they'll consider it a firm delivery commitment. Discussing project dates in the concept stage is okay as long as it's clear they're tentative, and there's an understanding dates will change after "discovery."

Twelve things needed to develop a schedule

1. Constraints imposed by stakeholders.
2. Availability of infrastructure (power, networks, space)
3. Embargo dates. (bank holidays, new years)
4. Competing projects.
5. Technology updates required before proceeding.
6. Resource availability.
7. Material factors. (equipment, etc.)
8. Environment/data analysis.
9. Dependencies.
10. Security constraints.
11. Task durations.
12. Maintenance windows.

Avoid establishing milestones and timelines before allowing the subject matter experts (SME's) to conduct the discovery.

Be Pragmatic

There's a tendency in sales to focus on positives but to avoid the negatives (i.e. realities) out of fear of blowing the deal. Most clients know projects don't always go 100% correctly. Along with the positives, be pragmatic about some of the challenges along the way.

Selling a rosy picture promising your solution without problems raises suspicion in your client's mind a) you are inexperienced, or b) you are not 100% honest with them. It's better to tell your client your experience in similar projects suggests there may be challenges, but your team knows the "land mines" and will be able to deal with them.

The client is interested in your services because they believe you bring the right combination of technology, resources and experience to deliver a successful solution. Keeping things realistic is part of becoming a trusted advisor.

Make Your Contract Understandable

A statement of work (SOW) should be written in a way quickly understood and absorbed by the reader. I've seen SOW's poorly written, making it hard to decipher what's to be delivered.

Seven key attributes of a good SOW

1. Keep all points in bullet form.
2. Clear scope of work.
3. Keep the number of pages to a minimum.
4. Divide the document into logical sections. (example: scope,deliverables,resources,budget,terms + conditions)
5. Avoid repeating the same content in different places on the document.
6. Provide a table showing measurable deliverables. (example: # servers to install)
7. Avoid marketing hype and excessive wordiness; this is a contract - not a sales brochure.

Don't Sell And Run

I've participated in many projects where the sales team mysteriously disappeared in a dust cloud after the purchase order was received. Conversely, I've enjoyed running others where the sales team remained engaged throughout the entire project life cycle. When the sales team remained involved, those projects always had the highest customer and team satisfaction.

I recommend salespeople continue to stay engaged after the purchase is signed, obviously to a lesser degree, but try to attend update meetings where possible and keep in the project loop. Staying involved demonstrates you care about its success.

Additionally, you'll be able to provide valuable insight because of your early pre-sale involvement. Staying engaged often brings hidden opportunities into view as project teams "peel back" the onion layers we spoke about earlier.

Sell What You Do Well

Avoid selling products or services outside of your organization's core business strengths, and purely for the sake of generating revenue. That approach loses money and ties up resources that could be applied elsewhere. When you can't deliver a decent product or service, it negatively impacts your company's reputation. Unless it's strategic, avoid selling outside your wheelhouse.

Good business focuses on marketing core competencies, products and services for a fair price at a decent profit margin. Vendors are entitled to a reasonable margin for providing quality products and services, trained resources, business stability, and ongoing product support

Never Knock Your Competition

One crucial lesson I've learned is never to speak ill of your competitors. There are no gains to slam your competition, and you'll never know when you may be working with them or for them in the future. Strategic partnerships abound these days, and I've been transitioned into multiple new employers through mergers and acquisitions during my career. Some of them were previously fierce competitors.

Most technology companies provide excellent products and services and have their own unique merits. Another reason to avoid slamming your competition is that your stakeholders may have deep brand loyalties to competitive brands. The fastest way to ruin a client relationship is to speak poorly of a company they respect. When asked to comment about competitors' products, try to start with a positive remark, and then follow it with a logical reason why your company would provide a better fit. This approach should apply not only to sales teams but also to all other vendor resources working on projects.

Learn About Project Management

Sales teams have comprehensive knowledge about the hardware and software products they sell. Still, when it comes to project management services, things can get a little foggy. Ask a project manager from your local office to run a seminar for your account team, covering what project management is all about. Learn about the key project phases, project management institutions, ITIL, methodologies, project structures, risks, staffing and budgets.   This information will help provide valuable insights and help with your credibility with your clients.

Get Your Back Office Team Involved

The "real world" is out in the field with your clients. Your sales admin staff usually sit behind a desk, and I'll bet you five bucks they've never seen your firms' products in a real-life setting, have never visited a client's site, and have no idea what your products do! Likely their only client exposure experience is from the information they glean while processing your sales orders through the system.

As a young inventory clerk working at head office, I recall the first time I visited a customer in the field. It was a life-changing experience. My entire perspective changed that day. It was a revelation to see the products in their natural habitat, and exciting to meet the customers who, before that point, only appeared to me as names on the order forms. I started forming better relationships with the customers and the field sales teams. Eventually, I was encouraged to take a job in sales, and ever since then, I've worked in the field.

Your admin staff is an integral part of your team, and they'll be more enthusiastic and capable when you give them a chance to get involved. One great way to introduce your back office staff into the field is to have them attend a client project kickoff meeting. Ask your customer if you could bring some of the back office staff onsite to meet them and get a tour of their facility. What better way to show the client all the people that work on their important account?

Key Points

☐ Involve a project manager and SME early.

☐ Avoid establishing timelines before discovery is done.

☐ Be pragmatic. The customer will appreciate it.

☐ Make contracts understandable.

☐ Stay involved in the project after the sale.

☐ Avoid selling products and services outside of your wheelhouse.

☐ Never slam your competition.

☐ Learn about project management.

☐ Get your back office team out into the field.

COMMUNICATING: THE ROOT OF MOST PROBLEMS

Since humans began working together in teams, good communications have been critical for successful outcomes. This chapter is a review of general communications skills as well as ideas on managing meetings and presentations. Good communications can help a project become a success, while poor communications are the root of most problems.

Keep The Message Crisp

When I worked at Data General, our country manager always asked us to keep communications "crisp" with our customers. It's challenging to present complex ideas in clear and simple ways, but simplicity should be your goal with general stakeholder communications. Avoid providing unnecessary information. Unless the communication requires complexity, keep it as simple as possible.

Listening / Responding

How we interact with stakeholders in verbal conversations is vital to the success of projects. We often take for granted the importance and power of active listening and also how we react and respond to questions.

Five essential listening tips

1. Take time to listen to what the other person is saying. People appreciate being heard and taken seriously. Look directly at the speaker and give your full attention.
2. Acknowledge that you hear and understand what the speaker is saying by nodding or giving verbal feedback.
3. When you don't understand something, ask the speaker to clarify their point.
4. Put down your smartphone and, unless you're in a web meeting, close your laptop. Holstering your devices shows you are giving 100% attention to the person speaking.
5. Write down key points. The reason to write things down is to record information, but taking written notes also sends a powerful visual message to the speaker. It shows you're focused on their words and feel them essential to record.

Two essential responding tips

1. Work at getting comfortable pausing before responding to questions. A pause gives you a chance to form a considerate answer and shows you feel the subject is worthy of thought. Pausing takes some practice because remaining silent is uncomfortable. Try counting to five before responding.
2. Before responding, repeat the other person's key points to show you've absorbed what was said. This step provides an opportunity for clarification and gives you time to think of a thoughtful reply.

Pick Up The Phone

Compared to e-mail, picking up the phone improve relationships, cuts through red tape and accelerates things. Without observing body language or hearing a voice in the conversation, an email's mood is at risk of being misinterpreted by the receiver of a message. If you have a sensitive topic, it's always best to call or, better yet, visit in-person. Don't forget to follow up on meaningful discussions with a confirming email.

Treat Your Email Like A Newspaper

Emails are so pervasive these days they are easily ignored. The subject line is your headline. Use the subject line effectively to get the attention of the reader.

Email Subject Headers

When changing topics or adding significant new points to an existing email thread, it's common sense to show a change in the email subject line to reflect the new theme. If the main topic hasn't changed, but further crucial information is needed, I'll append the original subject line with revised information inside angle brackets. For example, Subject: Project X install schedule <now April 12 - details enclosed>. If the topic of the thread is entirely new, you should send a fresh new message.

Before You Hit "Send"

Sometimes we get into situations that raise our frustration levels. Our first reaction is to hit back with a scathing email.   I remember times when project roadblocks caused me to shake my head in disbelief. The next thing I knew, I was composing an email containing my candid thoughts on the issue! Fortunately, most times, I let myself cool off and carefully read and edited my draft message again before hitting the "send" button.

During the heat of the moment, your message might look appropriate, but allow a cooling-off time before firing it off. Emotions get in the way of rational thinking. Think about how the receiver will interpret the message.   For significant issues, if possible, sleep on it overnight and look at your draft the next day. More than likely, you'll realize some editing is required – then hit send!

The Rule Of Five

In projects, we often have lots of valuable information that we want to share with our stakeholders. Unfortunately, because of the way our brains work, people have limited short-term memory. The limit of our audience's retention means we need to be careful how much information we throw them and which key messages we want them to remember.

A useful rule to adopt is the rule of "five things." If you have an essential message in an email, presentation or discussion, try and limit your key points to five things or fewer. In fact, the fewer the better. Your message will have a greater impact on your audience if it's easy to remember.

Anatomy Of A Good Status Report

Stakeholders always expect forward motion during projects. Unless we publish regular updates, stakeholders lose trust, and then their imaginations start to run wild. Until the final product is delivered, strengthen stakeholder confidence by posting frequent updates.

Because your stakeholders will be looking for progress indicators regularly, reports should have data displayed in a consistent format for quick and easy comparison to previous reports. This concept could be compared to an automobile dashboard. In a car, the speedometer and other gauges are always in the same place and use the same measurements so the driver can monitor changes at a glance during the journey

Eleven attributes of a good status report

1. Keep your report to one page, if possible.
2. Use a consistent title for each report.
3. Use a consistent formatting for metrics.
4. Consider a "dashboard approach".
5. Generate updates regularly.
6. Limit to critical points only.
7. Include progress, measurable metrics, next steps.
8. Include detailed delivery information for materials.
9. Describe major issues + impact on schedules.
10. Include the time for the next scheduled update.
11. Invite questions or feedback.

Guidelines For Effective Meetings

In recent years the number of meetings we attend has increased dramatically. There's a massive trend towards working outside the office, and thus remote tools such as Zoom, WebEx and Skype are prevalent. Regular meetings are essential in projects, but depending on how well you manage them, meetings can impact productivity both positively and negatively. In some circles, meetings are the only time when a project team can gather.

Unfortunately, as the number of meetings increases, the amount of time to accomplish productive work decreases. Excessive meetings impact your team's ability to do other work that could be crucial to the project. The more attendees you invite, the more difficult meetings become to schedule and manage.

Sixteen tips to make meetings productive

1. Keep the number of meetings to a minimum.
2. Share a clear meeting-agenda in advance.
3. Keep sessions brief and maintain focus.
4. Frequency of meetings depends on the situation:

a. Escalation issues: daily / hourly as needed.

b. Status meetings: weekly.

c. Executive cadence: monthly.

5. Keep the meeting attendees to only those required.

6. Invite subject matter experts if needed and have them speak on their area of expertise.

7. Use screen sharing as appropriate to engage your remote audience.

8. Consider the time zones of your attendees when scheduling.

9. Avoid scheduling meetings during meal times.

10. Establish ongoing meeting cadence at the kickoff.

11. Gatherings of larger groups should be treated as seminars, lectures or workshops.

12. Establish "work-only" periods for your teams, strictly avoiding meetings during those times.

13. Use the built-in IP voice links on meeting apps instead of phone lines.

14. Allow time for audience questions and inputs.

15. Meeting minutes should be published (if required) quickly following the meeting.

16. Record meetings only after getting acceptance from all attendees.

Video Meetings

In the age of remote meetings, the way you present yourself on video is essential. Think of yourself as the producer, director, set coordinator, camera operator and the star of a TV show. Carefully position cameras, so you're looking directly into the lens. Avoid camera angles making you appear disengaged or intimidating. Backgrounds should be neat and uncluttered, so things behind you won't distract your viewers. Good lighting is another critical consideration. All video meeting applications have a monitoring screen so you can preview how you look to your audience. Adjust the brightness in your room, so your facial features are clear and evenly illuminated. Unless you want to look like "Evil Doctor

Doom," avoid backlighting or uplighting.

Avoid Politics

Sharing political views should be avoided. Leave your political opinions at home. Think about the consequences of sharing your views publicly. What you post on social media could be accessed by clients and stakeholders.

Presentations

Project and technical presentations can be some of the driest, tedious and most boring sessions on the planet. To maintain audience attention, keep them actively involved. Ask for feedback, ideas, and contributions periodically during your presentation. Don't put up with people talking or looking at their electronic devices during your meetings. Expect complete attention from your audience.

Five essential presentation tips

1. Keep the number of slides/points to a minimum.
2. Use graphics to tell the story where possible.
3. Don't read from the slides. Use bullet points and your own words.
4. Get your audience to participate (feedback/questions).
5. Use a whiteboard where appropriate.

Your Audience

Who is your audience? When you think about it, each of us lives in a separate universe with different views, priorities, knowledge and backgrounds. Each member of your audience has a different perspective on things. When presenting ideas, think ahead about positioning your message to be easily grasped and understood so your audience can relate.

When No One Says A Word

Have you ever attended meetings where the attendees are silent? I've participated in meetings where the host invited the entire team for an important update, and when feedback was requested, no-one said a word! The silence is usually an indicator of a problem. Try and encourage honest feedback. There's often something wrong when people stop actively participating, and they become passive observers instead of active team members.

Following these types of meetings, it's an excellent practice to privately contact some of the attendees to get some honest feedback, find out if there's a problem, and if there is one, then fix it!

Meal Time

Lunch and dinner times should be avoided for meetings or presentations unless there's an urgent escalation. One of the few breaks we get during the day is for meals. Humans require food! I suggest letting people eat their food in peace. Be sure to take time zone differences into account.

Key Points

☐ Keep the message crisp.

☐ Listening and responding are vital skills.

☐ Pick up the phone.

☐ E-mail subject lines are like a newspaper headline.

☐ Think before you send that e-mail.

☐ The rule of five helps your audience remember.

☐ Give your stakeholders regular progress reports.

☐ Consider the anatomy of a good status report.

☐ Make your meetings brief, concise and compelling.

☐ Video meetings are your personal TV show.

☐ Avoid politics and be careful what you post on social media.

☐ When presenting, understand your audience and get them involved. Keep it simple.

☐ When your audience is silent, follow up privately to get feedback.

☐ Avoid scheduling meetings during mealtimes.

METHOD AND PROCESS: FRIEND OR ENEMY?

Without methodologies, processes and procedures, there would be chaos. These elements provide a vital structure to help guide us safely and efficiently to our objectives. Consistency is required, but tweaks could be needed to address unique project requirements. Methodologies, processes and procedures are like Russian nesting dolls. A methodology is a system of managing to achieve an objective. The inner shell contains the phase steps that define a logical sequence of activities. Nested inside that shell are the procedures that include detailed tasks to achieve a result.

You should always follow some type of methodology. A methodology provides structure and forms a framework for your plan. Without this, you'll be flying aimlessly without a rudder.

The following graphic illustrates a simple sequential methodology and shows how the phases, processes and procedures are nested:

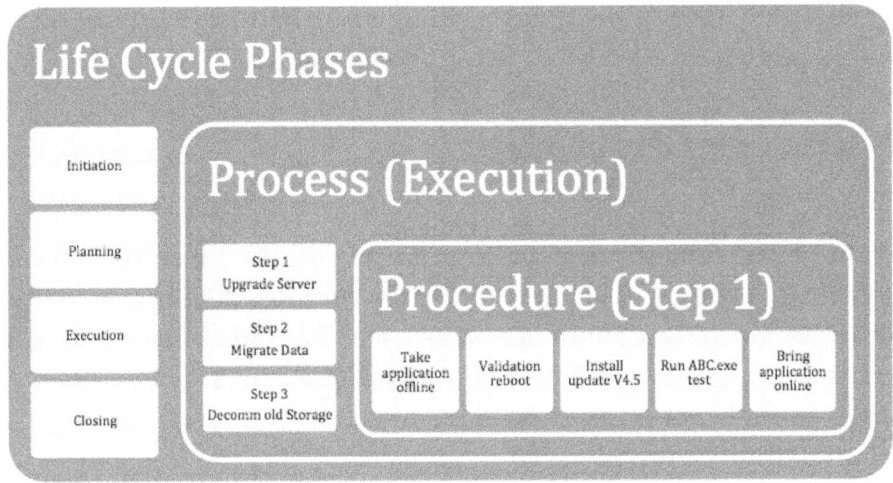

The Project Management Institute (PMI), PRINCE2 and other organizations have specific methodologies they sanction as standards for project management. Other methods include Agile, Six Sigma, Lean, Kanban, Event Chain, Critical Path Method, Extreme Programming, and Project Framework.

Exploring these in detail is outside of the scope of this book, but in a nutshell, each of these methodologies has its distinct advantages and applications. Some methodologies follow a strict sequence, while others promote more flexible methods to get things done.

The best methodologies and processes are those appropriate for the project. Avoid being forced into using the "method-of-the-month". If you're not already bound to one, pick the right methodology to suit your project.

Map It Out / Tell The Story

Map out the end-to-end process with your team. You don't need to get into detailed planning, but this step will create a dialogue that will help you map the core project process. Make sure you have the correct subject matter experts on hand. I recommend to do this on a white-board or using post-it-notes.

Engineering Procedures

Always follow processes and procedures developed by product engineering and support. These are a result of extensive (and expensive) testing and analysis during product development and must be considered mandatory. Deviation could compromise the project, and it's supportability. Procedures could be version-specific, so make sure the procedure matches the release level you plan to implement.

Data Center Change Tickets

The change control process can be one of the most frustrating things in IT. These processes are in place for excellent reasons, but they can be challenging when planning activities. Another aspect of change control procedures is that they are typically non-negotiable.

On any new project, <u>it's vital that project managers understand the client's change process</u> and carefully plan activities with that process in mind.

Non Essential Steps

Avoid mapping non-essential steps into a process. Such measures will bog-down and discourage your team members. Maybe it's my imagination, but sometimes it seems that people who get promoted to senior management like to cook-up ways to make work more complicated. Inefficient processes and procedures cause infinite grief on the ground, thanks to a lack of empathy by those who create them.

The Cost Of A Useless Process

Poorly conceived processes and unnecessary procedures cost organizations time and money. Managers have a habit of adding small tasks to existing workloads, and sometimes these little extras are a critical and valid requirement, but often they are not. An example could be asking all staff to prepare a special report on their activities every week. The task in itself may seem insignificant, but when added onto a pile of other small extras, the impact on resources becomes dramatic. It's like the proverbial last-straw attached to the back of the poor camel. These mini-tasks have a higher cost than is realized!

A simple formula for calculating such costs in an organization is to multiply TH (task hrs/week) x HC (hourly cost) x NE (number of employees) x NW (number of weeks). The following table is an example of the costs of executing a range of small tasks in an organization of 100 employees whose internal cost is an average of $50 per hour.

Hours / week per person to execute task	WEEKLY Hours / week 100 staff Cost $50/hr	MONTHLY Hours / month 100 staff Cost $50/hr	ANNUAL Hours / year 100 staff Cost $50/hr
.25 Hours	25 $1,250	100 $5,000	1,200 $60,000
.5 Hours	50 $2,500	200 $10,000	2,400 $120,000
1 Hour	100 $5,000	400 $20,000	4,800 $240,000

Asking a staff of 100 to spend an extra 15 minutes per week on a report may seem like a small request, but it costs the organization $60,000 per year!   Is the expected value to be derived from the report worth $60,000? By the way, the cost model doesn't consider the impact these mini-tasks have on productive activities. What other things could your staff do with this time if they gained it back?

It's important to remember the costs are additive. Every time a new small task is added, the costs continue to accumulate. Managers seem to forget about all the little things they ask you to add to your workload, and things pile up quickly. Consider the cost of unnecessary tasks before adding them to the team workload.

The Bean Counters

Project accountants are the project manager's friend. Accounting plays an essential and necessary part of business, and measurement is an integral part of project management.

Where accounting and measurement wander from common sense, we find the bean counters. The bean counter's role is to track nitpicky metrics that make no sense and have no impact on anything except perhaps someone's bonus. Their priority is for everyone to comply with their relentlessly, irritating demands, no matter how counterproductive. A classic tactic of bean counters is to intimidate their victims by stack-ranking them compared with their peers, then publishing the results for all to see. While these reports are politely reviewed by those targeted, you can hear the sound of grinding teeth if you listen carefully.

I agree measurements are essential to track progress, but when unnecessary measures obstruct real work, they become barriers. This topic goes hand-in-hand with the costs of an inefficient process. Consider the impact bean-counting tactics could have on morale and productivity.

The Latest Fad

Every several years, a new fad, management strategy or method-ology emerges to try to reshape organizations. Some examples of strategies I've encountered are "Quality circles," "TQM - Total Quality Management," "Search for Excellence," "Quality is Free," "Six-Sigma" and "Juran - Quality Improvement." I still have many of the textbooks sitting on my office bookshelf. In every case, my employer enthusiastically embraced these approaches but si-lently abandoned them a short time later.

Why did my employer introduce them in the first place? The an-swer is they viewed these as quick and easy ways to fix perform-ance and quality problems. The hope was a new approach would fix the issues. The common problem was that applying the ap-proach added more workload to resources, and often made things worse. In the end, it seems like those who benefited the most were the consultants who charged our company vast sums of money.

There's no question that each of the above theories has positive benefits when applied as designed. I learned many useful things through the training, and some things I still use to this day.

Ask these eight questions before adopting a new methodology

1. What is the new approach trying to resolve? You should understand the root cause of the problem before trying to fix it.
2. Financially what will it cost / how much will it save?
3. Are there any other alternatives, if so, what are they?
4. Do your staff have the workload capacity and motivation to adopt the new approach?
5. How long will it take?
6. Is your organization committed to the new approach as a

long-term solution? If so, what measures are needed to maintain interest and enforce compliance?

7. Are there "champions" in your organization that can mentor others? Will they have the time and capacity to do so effectively? How many will you need?

8. How will you measure its effectiveness?

Key Points

☐ Methodologies come in different flavours. All projects should follow a methodology.
☐ Use the appropriate methodology for the project.
☐ Map out the process with your team in the early stages.
☐ Always follow procedures established by engineering.
☐ Understand the change ticket process.
☐ Avoid building non-essential steps into a process.
☐ Be aware of the additive financial and human cost of tasks added to staff workloads.
☐ Consider the cost of bean-counting tactics.
☐ Carefully assess the cost of committing to a new fad approach before diving in.

PLANNING: THE PROCESS OF MAPPING IT OUT

What's the plan? How long will it take? What's the schedule? These are probably the three most common questions asked of project managers. Stakeholders often asked me these questions at project kickoff meetings, but I always found it a little surprising that stakeholders expected a completed plan the same day the project started. Preparing a plan usually takes time and effort, but despite being unrealistic, these kinds of questions offer a great opportunity for the project manager to stand up and walk stakeholders through the whole process. This chapter focuses on tips related to project planning.

The Seeds Of Failure Are Planted Early

The title says it all. If projects are scoped or planned poorly from the outset, they are doomed to fail. The best way to avoid this is to be an active participant in the project concept/initiation phase. That way, you have a role in influencing the project scope and early planning.

Too often, the project manager is brought into the project after the purchase order and contracts are signed. Sadly, by that time, the poor PM has little recourse but to pick up the pieces, start from scratch, or reluctantly lead a botched plan. How do you influence senior management to get you involved upfront? The answer is to suggest they buy a copy of this book and read the second chapter.

Software

Use a planning software tool that's appropriate for the job. You don't necessarily need MS-Project to create a project plan. Projects can be effectively planned on spreadsheets if required. The advantage of using MS-Excel for project planning is that it's a very flexible tool, and because many stakeholders use Exel, your planning documents are easily shared with others. On the other hand, specialized project planning tools such as Microsoft Project have built-in features that can accelerate your planning process.

See It In Person

Nothing can replace the power of firsthand observation. Seeing the physical hardware gives you a better perspective and gets the picture into your mind. It has a mental impact, provides context and expands your thinking related to the project. Sometimes it's not realistic or even possible for everyone to "see it in person." In this case, the next best thing is to provide a photograph or model.

Some years ago, I attended a steering committee meeting at a hospital in Toronto where a large-scale SAN migration was to take place. The committee included members of the hospital administration and health care professionals. Despite our repeated attempts to map the project steps on the white-board, the non-technical committee couldn't grasp the project concept. After the meeting, I went home discouraged and tried to figure out how we could explain a complex technical project to a group of non-technical folks.

That evening watching my young son Harrison play with his LEGO bricks, I had an inspired idea. What if I used the bricks to show the board how the project would be executed? I pulled a pile of bricks from his toy-bin and started building a toy model of the hospital data center, complete with coloured blocks representing the project's components.

I arrived at the next board meeting with a large paper bag that aroused some curious looks from the attendees. As I pulled my data center model from the bag, I thought the Hospital's IT manager was going to fall off his chair. Running through the project steps by moving the models' parts like chess-pieces, the colourful three-dimensional model successfully demonstrated the deployment concepts to everyone's delight. It was a risky move, but we had some fun, and they finally understood the projects' logistics.

Get-There-Itis

I have a private pilot's license. At flight school, I discovered aviation has many similarities to project management. Risk management and planning are vital in the aviation world. One of the things we learned in ground school is that many airplane crashes are rooted in a push to get to a destination despite known risks. This deadly compulsion is known as "get-there-itis." The situation usually begins with the pilot feeling pressure to stick to a schedule while ignoring risks such as bad weather and low fuel. This pressure can be overwhelming. Thankfully most pilots delay their departure until it's safe, but some yield to the temptation, often resulting in a fatal accident.

How many projects do you know which have "get-there-itis"? In the project world, "get-there-itis" is a project that places the schedule before anything else. How do these dates get picked in the first place? More often than not, the timing is rooted in a business goal without consideration for its planning. In IT, marching forward despite known risks usually won't create fatalities, but it can cause problems resulting in negative impacts to the business, such as outages and data loss.

The End User Vs. End Client?

The end-user is the one who directly interfaces or benefits from the product or services you've deployed. You may not always have access to the end-user. For example, returning to our theoretical project X from chapter one, the *end-user* is the bank customer who accesses an ATM to withdraw money, but the end client is the software development manager. The *end client* is the person who defines the requirements and specifications for the final product.

A great question to ask during a project kickoff is, "who is the end-user and who is the end client (who represents their requirements)?" This question can help identify who will have input to the final product and how it should look. It's advisable to always engage with the end client in the early stages to understand their requirements and specifications and confirm their expectations. Continue to engage the end client while the project develops to obtain their feedback.

What Is Quality?

Quality is not necessarily the same as excellence or robustness. Quality is delivering a product or service meeting a specific requirement or design specification. As an example, if you're building a product designed to fail after a single use, and it does so, then the product successfully meets the defined quality requirements. Building or over-delivering a product or service exceeding design requirements and specifications creates waste unless exceeding expected quality is part of a strategy.

Plan B

Contingency planning is a crucial project risk mitigation component. Have a "plan b" to protect yourself from unexpected circumstances. Always have someone designated to backup your critical resources as a contingency because you may need to cover for vacations, sickness, or even an exit from your company. Ask your team, "what happens if...".

Infrastructure Takes Longer Than Expected

Networks, power, and space seem to be significant challenges in the 21st century, usually in that order. In delivering projects, many delays result from the time it takes to deploy infrastructure and all the paperwork required to put it in place. Network planning should always be validated as networks require multiple levels of planning (IPs / physical/configuration). If errors occur during the planning stage, it takes additional time to correct the problem.

Draw A Picture

The adage "a picture is worth a thousand words" rings true in IT projects. Draw a picture of your project. Most projects usually have a physical layer (visible structure) and a logical layer (sequence/flow/streams). As simple as it sounds, mapping your project into a graphic will be challenging but very rewarding. A drawing thrown up during a meeting or planning session will encourage far more participation from your audience than a narrative.

One day, a desperate sales team asked me to chair a project kickoff meeting with little advance notice. In the small amount of time I had, I quickly created a simple project map showing the physical devices and their deployment sequence. We spent over one hour reviewing that single slide. Sometimes a simple picture is all that's needed to kick start a great planning session.

Ask Stupid Questions

At planning meetings, some IT professionals hesitate to ask questions for fear they may look incompetent in front of their peers. Sometimes the best questions are the ones we think are stupid. I once attended a Telco planning meeting where the facilities team was waist-deep in planning the network infrastructure for a storage subsystem. I raised my hand to ask my "stupid question." When I asked if space had been set aside for future expansion cabinets, I was surprised to find it was a question no-one had considered. My question triggered a re-design and helped save a future headache and tens of thousands of dollars. Don't be afraid to ask stupid questions

Split Work Into Chunks

Tackling a project can be overwhelming. Often if you split the work into manageable chunks, it becomes much easier to grasp, manage, and report. Manageable chunks are much easier to delegate.

Information Security

Information security and access is a sensitive topic. Spreadsheets containing sensitive data such as IP addresses, contacts, server names, and passwords may be needed for IT planning and work. Still, they must strictly be secured to avoid unauthorized parties' access.

Key Points

☐ Projects usually fail due to poor early planning.

☐ Seeing it in person brings a better perspective for planning.

☐ "Get-there-itis" can be deadly. Placing the schedule before anything else puts the project's success at risk.

☐ Ask who is the end-user and the end-client.

☐ Quality is delivering against a requirement.

☐ Always have a "plan-b" in your back pocket.

☐ Power, networks and space always take longer than expected to prepare.

☐ Draw a picture to engage your team during planning sessions.

☐ Never be afraid to ask stupid questions.

☐ Split work into manageable chunks.

☐ Be very careful with sensitive data in your possession.

MANAGING AND LEADERSHIP

Managing a project is much more than merely organizing and scheduling people. Effective leadership is required. Leadership is the art of managing without controlling, instilling confidence, taking ownership, sharing the credit for successes, and motivating your team by genuinely helping them succeed. Team motivation makes the difference between a "satisfactory project" and an "outstanding project."

Motivated team members are inclined to work with less direction, have higher interest and engagement in their work, and usually deliver better quality results. Motivated teams are less of a burden on management and increase stakeholder satisfaction.

Motivation is generally related to rewards and is usually financial, career-oriented or emotional. Each individual has a key motivator, and it's an art to be able to identify motivators and apply them to get the best out of individuals. Other motivation factors include providing a pleasant environment and the right tools to do the work.

Run The Project Like It Was Your Company

Try and run your project like it was your own company. After all, a project is a mini organization. Each project has a leader, staff and a product or service to deliver to customers. You are the CEO, the project teams are the employees and the managers above you are the board of directors.

Sit On The Fence

The project manager must sit on the fence. The PM is the client's advocate but may need to represent stakeholders with different interests. It's a tricky balancing act. At times, this can be an uncomfortable position, and it takes sensitivity and diplomacy to navigate challenges if they materialize. Keep an open mind and try and see the view from both sides of the fence.

Are You A Manager Or A Controller?

There's a difference between managing and controlling. A good manager manages people, and a lousy one controls people. Managing people requires skills such as coaching, mentoring, and encouragement. Managing provides guidelines for success and uses individuals' talent and experience to achieve a positive result.

Controlling people is an attempt to impose one's will on subordinates. Controlling requires few skills except for bullying and intimidation. Because no one likes to be a "puppet," this ultimately leads to the team's failure and, eventually, the demise of the controller leading it.

Micro Managers

Give none of your valuable time to micromanagers. The micromanagers are the ones who have no idea what's going on, they're probably hiding their lack of skill, and their only purpose will be to get in the way of the real work upon which you focus.

Ask Your Team For Their Opinions

The project manager is not a "superior being"; she or he doesn't know everything. A good manager leverages the knowledge and experience of the resources on the team. Ask your team for their opinions and genuinely listen to them.

Staffing With People Who Failed

People learn from mistakes, and they rarely repeat serious ones. I once managed a large project where a senior resource made a significant error, despite being one of the best specialists in the company. Eventually, the error was corrected, but the angry customer directed me to take him off the project, worried he would make the same mistake again.

I pushed back and explained that he was now the very best person we could ever have on the project. Why? He was the best person because he learned from the failure, and he was the person most unlikely to make the same error, and any more like it, again. The project ended successfully with the same resource remaining on board. The best part of the story is that the client asked for the same person to return on future projects.

Workload

It's vital to understand your staff's workload ratio. "Workload" is a measurement of activities to be delivered within the time available. Your resources should have sufficient time to conduct their core tasks without compromising the quality of work, but also they need an extra 20% of the runway.

What should staff do with the extra 20%? This time could be

wisely-used for customer relationship-building activities, training and an escalation reserve. Should a resource need to respond to an escalation, they'll have available time to change focus and attend to the problem until it's resolved.

Some organizations expect their staff to be at 100% capacity. I suggest 80% capacity should be considered the maximum, and beyond that, managers should consider assigning work to others with available cycles.

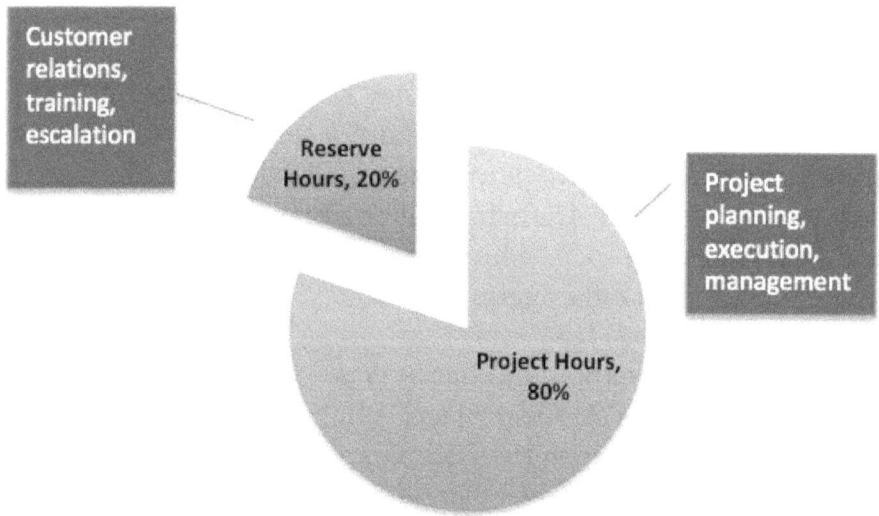

Recommendation: Work should be divided into 80% project work + 20% reserve for customer relations / training / escalation.

Responsibility For Your Staff Welfare.

Permitting resources to work under high workload conditions and extended periods without rest is placing that person's health at risk and creates a liability for your company. Such conditions cause employee burnout, which can translate to poor delivery to your customers, reduced productivity, project errors, health issues, loss of good people and potential litigation.   If you have IT resources working onsite for extended hours, make sure they can check into a local hotel after their shift. Their welfare should be paramount, and we don't want tired and dreary eyed staff driving the roads placing themselves and others in danger. Managers must consider the consequences and develop programs to identify and relieve stress and excessive workloads. Field staff should have pre-approved options to book into a hotel if working late or long hours away from home.

Monitoring Morale

Morale is like the heartbeat of your team. With robust and positive morale, you're more likely to have a positive and efficient work environment. People will be happier and work together better, and they'll be able to handle peaks and valleys in stride. Strong morale also translates to positive interactions with stakeholders and better customer satisfaction. When weak or negative morale infects your team, you will find it harder to manage, and the team could become dysfunctional. Their quality of work and productivity could suffer and impact your project negatively. It's essential to monitor your team members' morale, and often it's not easy to gauge how people are feeling accurately.

An excellent approach to assessing morale is a weekly "one-on-one" meeting with individual team members to get their candid feedback on things. This meeting is typically thirty-minutes long and is an opportunity for the team member to privately provide an update on their workload and challenges they face. Because the session is between the manager and a team member, there's a more significant opportunity for a candid conversation than in a group discussion. For this to work correctly, team members must be open and honest, and managers should listen carefully and take the appropriate action with follow-up. This approach will work in formal and informal reporting structures (e.g. manager-employee or project lead-team member).

Recognition

Publically recognize good work.   Awards handed out privately are beautiful but take the opportunity to make them a special event. When someone does something exceptional, shout it out and share the news with everyone.   Recognition is not only for your employees and team members. If a client does something noteworthy, there's no better way to reinforce a partnership than to recognize those outside your organization.

I recall one occasion where I was working in an IT project at a local hospital. A team member on the client's staff provided out-standing technical support during a data migration event. With all the stakeholders attending, I presented a special award to the individual. This gesture of recognition had a powerful positive impact and appreciated by the recipient and the client.

Criticism

While recognition should be carried out publically, criticism should be in private. No one likes to be criticized. Although criticism is a way to highlight things that need improvement, the way it's delivered is critical. To simply point-out fault does little but provide discouragement. The bitter pill of criticism always should have some honey to wash it down. Administer criticism with a healthy dose of encouragement.

Spontaneous Appreciation

Making someone feel appreciated is one of the most powerful things you can do. If there's one thing you should remember from this book, remember this tip. It will provide huge rewards. Once in awhile, stop and hand-out some sincere out-of-the-blue appre-ciation. When you see someone doing a good job, stop and say, "Hey, I just wanted to tell you I like how you handled that!". This conversation takes very little time and is a compelling way to build relationships. You'll brighten someone's day and will feel great for doing it!

Why Use Subject Matter Experts?

A subject matter expert (SME) is a person with expertise in a specific area of knowledge. In IT, examples are expertise in particular types of databases, deployment best practices, or different migration methodologies. SMEs are valuable through all project phases, and you should have them involved starting from the initiation stage. They can offer insights into best-practices and risks. Their knowledge can make a significant difference in the outcomes of projects. If you don't have the required SME on staff, it could be prudent to hire one outside your organization.

Six reasons to use subject matter experts

1. It's a sensible resource management practice.
2. It places responsibility for inputs with the appropriate resources.
3. SMEs are formally trained and have privileged access to support systems and information others don't have.
4. It reduces risk.
5. It reduces costs by avoiding project land-mines.
6. It improves stakeholder confidence.

A Little Knowledge Can Be Dangerous!

There's no doubt that project managers are incredibly bright folks, but sometimes they're tempted to show off their technical knowledge and offer advice when they shouldn't. The adage, "A little knowledge can be dangerous" applies perfectly here! Project managers don't always have access to the latest training and information. Despite having good intentions, they could lead their client into a bad situation. PMs should always use the appropriate resources to secure and deliver technical information to stake-holders.

Key Points

☐ Teams are motivated by leadership. Lead and help your team succeed.

☐ Run the project like it was your company.

☐ Sit on the fence, see different perspectives.

☐ Be a manager, not a controller.

☐ Solicit your team for their opinions.

☐ People who have failed could be your "star" resources.

☐ Understand your staff's workload. Leave a 20% runway.

☐ Monitor morale. It's the heartbeat of your team.

☐ Publically recognize good work, privately criticize.

☐ Practice spontaneous appreciation.

☐ Use your subject matter experts effectively.

☐ A little knowledge is dangerous. Project managers should stick to their role and <u>not offer technical advice.</u>

EXECUTING

As managers of projects and people, we all have challenges with execution. We usually have a single chance at it, and we often have a lot of eyes on us while we're doing it. Successful execution is contingent on the planning that comes before it, effective communications, and of course, precision in delivery.

Pre-Flight Inspection

Before every takeoff, pilots conduct a walk-around inspection of their aircraft to make sure things are secure and safe. The walk-around, or pre-flight inspection, is mandatory and always performed in a specific sequence. Pilots make no assumptions and they conduct every inspection as if the condition of the aircraft is unknown. There's a specific pre-flight sequence for each type and model of aircraft.

Before commencing any project activity, I recommend conducting a pre-flight review with your team. This review assures that everyone is on the same page.

Ten things to check before implementing

1. The objective of the activity.

2. A detailed task plan.
3. Architecture documents completed/shared.
4. Software licenses secured.
5. Change tickets approved.
6. Pre-work completed.
7. Checkpoint times/conference bridges.
8. Start and finish times.
9. Resources assigned.
10. Escalation process.

The type of activity will dictate the content and level of detail for your inspection. In my projects, I conduct these meetings 24 hours in advance to allow for last-minute changes if required and then do a final check immediately before the activity starts.

Watch Your Fuel Gauge

In today's fast-paced IT environment, capacity always runs out faster than planned. By capacity, I'm referring to networks, switching, storage, servers. Keep an eye on capacity planning, and forecast 4 to 6 months in advance to account for lead times for upgrades and replacement hardware.

In my experience, I've observed repeated panic situations where corporations have run out of physical capacity and are at risk of not being able to run backups or store data. This problem is preventable. Do your planning and measure at regular intervals along the way. One great idea is to set up a periodic alert in your email system to remind you to check your fuel gauge.

I also want to mention software licenses. They are crucial but forgotten until something stops working. Expiring licenses can bite you in the rear without prior warning. Keep a database of their limitations, expiry dates and assign a team member to manage this information. When planing any implementation, always add "licenses" as a task item.

Stay A Step Ahead

Always prepare one or two steps ahead of the next activity. Lining up materials and resources ahead of the required schedule is a good practice. Looking ahead can highlight unknown challenges, and gives you some runway to correct things should they not be as expected.

Good Enough Versus Perfection

Often the right thing to do is to release a product that's "good enough" versus holding something up to get perfection. Understand the requirements and explore any opportunities to release a solution or product in iterative stages. This strategy could allow you to accelerate getting a solution into the hands of the end-user ahead of time and then execute minor changes to achieve the final state.

Key Points

☐ Excellent execution is contingent on proper planning.
☐ Do your pre-flight inspections, see the checklist.
☐ Always look one or two steps ahead.
☐ Monitor the environment (physical and licences).
☐ Sometimes good-enough vs perfection is a valid approach.

BARRIERS AND HOW TO HANDLE THEM

B arriers exist in life, and they happen all the time. Some are major, and some are minor. Most are a pain in the butt. In IT projects, a barrier could be a technical issue, contract, resource, policy, regulation, procedure or a stakeholder.

Five ways to deal with barriers

1. Comply - meet the requirements.
2. Avoid - change your course and go around.
3. Compromise - negotiate / change rules.
4. Ignore - go through and accept the consequences.
5. Abandon - halt the execution.

Barriers and persistence go hand-in-hand. Be persistent, and never let a barrier stop you. As a last resort, abandoning is a viable option, and sometimes it's the right thing to do.

Another option is to ignore the barrier and proceed straight ahead. Should you decide to ignore a project barrier, you must be prepared to accept the consequences. If you are running a project, make sure you document any deviations from the plan in the change control system.

Raise A Little Hell

If something doesn't look right or work right, raise a little hell. Innovation and improvement rarely are achieved by following the status quo. Challenge your management team. If they're worth their salt, they will listen. No one knows the situation better than people on the ground. When challenging the status-quo, always include a <u>proposed solution</u> to the problem along with the challenge.

Saboteurs

In some projects, stakeholders try and sabotage things. Sabotage comes in many forms and not always based on evil intentions. The most common type of project sabotage is resistance to co-operation. The resistance causes project delays and all kinds of mayhem.

Nine reasons stakeholders will sabotage your project

1. They don't buy into the project objectives.
2. They're aware of a risk that has eluded the project team.
3. The project is poorly planned, and they envision disaster.
4. They don't like the platform.
5. They don't like the project manager.
6. They aren't trained or prepared for a new solution.
7. They have a fear of new technology.
8. They weren't consulted.
9. They don't have work cycles to participate in the project.

Once a stakeholder becomes a saboteur, a lot of effort is required to move the project forward. The best way to prevent sabotage is to involve stakeholders upfront during the project concept/initiation phase. As stated several times in this book, engaging those impacted during the initial stages is proper project management.

The Gatekeepers

I was once in a project where there was a "gatekeeper." This gatekeeper's job was to review compliance and authorize the deployment of new technology. The gatekeeper had such stringent requirements; it was almost impossible to get anything deployed in the environment. The meticulous rules made us all a little insane.

Gatekeepers are usually in place for an excellent reason. Despite your frustration, you must always treat the gatekeepers with patience, respect and courtesy, because they often have the power to tighten their grip on things if they don't like you. Try to look at things from the gatekeeper's point of view. Ensure you clearly understand their requirements at the outset to comply with the rules they're enforcing.

Sometimes you need to escalate to get around the gatekeepers; if so, make sure it doesn't become "personal." You don't want to upset the gatekeeper!

Key Points

☐ There are five ways to deal with barriers (comply, avoid, compromise, ignore, abandon).
☐ Be persistent.
☐ Don't be afraid to raise some hell. Call-out a problem when you see it, but always include a <u>suggested solution</u>.
☐ Prevent sabotage by being inclusive.
☐ Respect the gatekeepers, understand and follow their process.

PROBLEMS: OPPORTUNITIES IN DISGUISE

Problems are opportunities in disguise. During my 20 years at EMC Corporation, I learned this is true. I'm not saying problems plagued EMC. I was proud to be a part of the fantastic people and technology offered there, but in any large deployments, you are bound to run into a few glitches along the way. Despite feeling the intense pressure when a problem manifested itself, things somehow always turned out positive.

In many cases, an issue highlighted a genuine customer need and, in the end, resulting in more business. In other cases, the lessons learned were used to improve quality and added to the support knowledge base. The challenge is to keep "cool" during an incident, communicate effectively, and work through the problem to its conclusion.

The Best Phase To Solve Problems

The best phase to solve project problems is any time before execution, the earlier, the better. Issues are always cheaper to fix in the initiation or planning stages than solving them after the project is underway.

Problem Management Plan

A documented problem management plan is vital to have in place.

Four things problem management plans should include

1. The way incidents are communicated.
2. Who is responsible for managing the problem.
3. How things are resolved, reviewed and documented.
4. Process for lessons learned / Root Cause Analysis (RCA).

Getting Senior Management Attention

Sometimes, you need to shake up senior management to get their support. The best way to get management's attention and focus is to speak in their language. Express a project impact in ways to which they can relate. Make sure reported effects are measurable and quantify-able. For example, if management is revenue-focused, quantify the risk in currency. (e.g. this problem costs $45k in lost production). Focus on the metrics that will grab the attention of your audience.

Get Tight With Your Support Organization

In IT projects, having a close relationship with the support organization is one of the most powerful assets you can have. A great way to develop this relationship is to meet with your local support manager and ask them to tell you about their team and show you how their support process works. Most IT support organizations are very structured and have processes you will need to understand and follow. Always follow the designated support process; if you try and circumvent the process, you will quickly get into the "bad books." Working in technical support is one of the most stressful jobs out there, and often the role is under-appreciated and taken for granted by clients. When someone from support resolves a critical technical issue for you, be sure to show genuine appreciation. A phone call or email of thanks to senior support management is always appreciated.

Technical Problems And How To Handle Them

Be quick and transparent about technical problems of any kind, including unexpected failures or human errors. Make sure you notify your client/sponsor about any issues as soon as possible after they happen. Problems happen in projects. How you handle issues defines a lot about your character and forms a base for trust and credibility with your client. When encountering a problem, I've found that technical deployment resources are so focused on trying to fix things that they fail to ask for outside help. They make heroic efforts to attempt to resolve matters on their own. These actions are admirable, but not a good practice.

When running into technical problems, I ask my team always to open a support ticket immediately to get other people involved. Opening a support ticket gets multiple people working the problem and provides a repository for log-files and dumps.

It's vital for the project manager to alert the sponsor about the issue and that the team is working on it. It's guaranteed the sponsor will find out about it, and it's much better that you call them before they call you! The account team (if you're a vendor) will want you to have an internal strategy call with them before calling the client. No doubt, the account team must be informed, but I always suggest communicating with the customer first because the problem could impact their business. Delaying alerting the customer so you can first speak to the account team could be costly. The longer you take to tell the customer, the wider the credibility gap gets, and the more time wasted before a fix is in place.

The last step is to monitor progress and provide regular status updates.

Problem Flow

Handling Failures And Faults

Whenever a failure or fault happens, it's always good to keep an open mind. Vendors are often quick to defend themselves, but they can be wrong. Regardless of your role in the project, don't jump to immediate conclusions before examining the data. Often the answer is the most obvious one. Look into the possibility that your team could have made an error. Find out if there were any parallel activities taking place during the fault. Be open and honest about failures and quick to accept responsibility.

Ahead of starting critical activities take a "before" configuration snapshot, if possible, so you have a baseline picture as a reference should you encounter a problem.

Failures Are Opportunities

As mentioned in the introduction to this chapter; failures are opportunities. It's painful to be a project manager during an incident where something has blown-up, but failures sometimes highlight infrastructure or process weaknesses. When they do happen, look for opportunities in failures.

Avoid "Shot-Gunning"

"Shot-gunning" is a method of trying to fix a problem by throwing all kinds of solutions at it "helter-skelter." When working through a technical issue that could have multiple fixes, deploy each one sequentially, and measure each result before moving to the next.

Systematic remediation is the recommended approach because if you execute a bunch of fixes simultaneously, you'll never know which one solved it.

Always start the troubleshooting process by establishing that the solution is in a currently supported baseline configuration and then proceed through the resolution process. Make sure the appropriate technical resources are engaged, and that a case with the support organization has been created.

Root Cause Analysis

When you have a severe failure, an RCA (Root Cause Analysis) report is an excellent thing to put together, regardless of whether it's requested.

Ten things a root-cause analysis report should contain

1. Support ticket numbers.
2. Equipment info and software code levels.
3. A problem description and impact statement.
4. Timeline of events.
5. Only factual information without bias.
6. Doesn't jump to quick conclusions.
7. Technical inputs written by qualified resources.
8. A logical/scientific analysis and conclusion.
9. Evidence from log files/dumps if available.
10. A resolution plan for future avoidance.

Persistence

Be persistent. If you believe in something strongly enough, keep moving forward, and don't give up just because you seem to be the only one on-board with a new idea or a way to solve a problem.

With any new approach, the innovators are usually alone, but innovation is the root of all progress, and you'll be in good company. Think about the great innovators; Galileo, Leonard Da Vinci, Henry Ford, Steve Jobs.

Look for allies to help support your beliefs and expand your circle of support. Come up with appealing ways to communicate your ideas to your audience and make the light bulbs on top of the naysayers' heads illuminate.

Cover-Ups

If you try and hide something terrible, it will always be uncovered. Your stakeholders will invariably find out if you're fibbing. Honesty is the best policy, it may be painful at times, but it's easier to be transparent. You'll build valuable credibility in the process, and you should never get punished for being honest.

Pick Your Fights Carefully

Fight for things that matter. Avoid battles over things that don't make a difference. Save your energy and focus on things that count.

Key Points

☐ Problems are opportunities.

☐ Problems are best solved in the planning phase.

☐ Develop a problem management plan.

☐ Get senior management's attention by speaking their language.

☐ Get close to your support organization.

☐ Be quick and transparent when encountering technical problems and keep an open mind.

☐ Avoid "shot-gunning."

☐ Create RCAs for severe issues.

☐ Be persistent.

☐ Never try and cover-up a problem.

☐ Pick your battles wisely.

CLOSING: AFTER THE CIRCUS LEAVES TOWN

C losing is often the most satisfying part of the project. It is the phase where all scheduled work is complete, and resources are released to work on other assignments. Yet, the project closing phase is an opportunity often missed when the stakeholders' focus quickly changes to new initiatives. Before closing your next project, consider the tips in this chapter.

After The Circus Leaves Town

A colleague of mine is a seasoned IT manager for a Toronto based legal organization. He observed there's always a lot of activity and excitement when the circus rolls into town, but the place becomes a "ghost town" when the vendor leaves. Make sure you follow up with the client post project to check the solution is working as expected and if any post-implementation support is needed.

Administrative Closure

Administrative closure is the process of getting required signoffs, releasing resources, and completing project paperwork. Before you reach this step, make sure you review project deliverables with your team and the client.

Eight steps of the closure phase

1. Verify deliverables are completed against the plan.
2. Project closure meeting.
3. Review lessons learned.
4. Verify deliverables with stakeholders.
5. Review the post-install support process.
6. Obtain signoffs, submit paperwork to project accountant.
7. Release resources.
8. Plan a follow-up call with the client.

Celebrate

A team celebration of what's been accomplished is important. Gather the project stakeholders together and recognize team members for their hard work and dedication.

Surveys

Recently, I visited my local car dealer for service. At the checkout, the cashier told me I'd be receiving a call from the manufacturer to rate my service experience. I was cautioned if I reported that my experience was anything less than "excellent," my service advisor, Mike, would receive a failing grade. All of a sudden, the survey became personal. Would Mike be punished if I gave some candid feedback for improvement? I felt like I was being intimidated into providing a good rating. This puzzle leads to the question, how honest are survey results?

Be careful designing and rolling-out surveys. Surveys should create honest and unbiased feedback and foster continuous improvement. I don't think it's fair to "prep" the customer before a review. Surveys have little value if there is pressure to influence the response. I also question surveys where bonuses are tied with

positive results or those where people will be punished for negative feedback.

Key Points

☐ The flight isn't over until you've landed and parked.

☐ Follow up with the client after the circus leaves town.

☐ Administrative closure is vital. Be sure to review deliverables, lessons-learned and submit any related paperwork.

☐ Carefully consider your approach to post-project surveys.

OFFSHORE FOLKS: TREAT THEM NICE

For the past 20 years, technology companies have increasingly been taking advantage of opportunities to increase efficiency and reduce costs by delivering services from overseas locations. Some organizations have outsourced their internal services to outside IT companies, while others have built extensive offshore facilities to serve their customers. Writing this book from a North American perspective, I'll use the term "offshore" to describe facilities in foreign countries delivering remote services. This chapter looks at the hurdles faced by offshore teams providing services to their customers.

Several years ago, I travelled to Bangalore, India, and visited the EMC facilities to meet a group of project managers who worked remotely with customers in North America. I found the PM's in Bangalore to be a fantastic group of young, enthusiastic people, and they made me feel a bit like a celebrity because of their keen interest in what I had to say. One of the most active topics we discussed was managing projects remotely and the related difficulties the offshore teams faced. As a local resource, I've delivered hundreds of projects to customers in the US and Canada with few issues, but the offshore folks have a much larger struggle to perform the same work. Let's take a look at these:

Why Offshore?

For a few decades now, organizations have been outsourcing services offshore to reduce costs. Publically this strategy is almost always positioned as a way to improve the customer experience. Most likely, the real reason is to reduce overhead costs and increase profit. In some cases, however, it's the only way companies can compete in a price-sensitive market. There are advantages, disadvantages, and tradeoffs.

Resistance

There's often resistance by North American customers to work with folks located offshore. Customers almost always prefer working with local resources. In some cases, however, the price and type of services purchased define the delivery method, and customers have no choice. Sometimes this translates into a frustrated customer, and everyone can sense the stress in the situation.

What factor makes customers prefer to receive services from local resources instead of offshore? The answer is "comfort-level." Customers simply feel more comfortable with resources they perceive to be physically closer-at-hand and in their time zone.

The Offshore Battle

In new projects, offshore resources face the immediate struggle of developing a relationship with a client thousands of miles away. The key to developing successful remote relationships with stakeholders starts with earning their trust. The challenge for offshore resources is that it's much harder for them to build stakeholder trust than their local colleagues. The reason for this is because it's not a level field.

Assuming they have an equal workload compared to the local team, offshore folks must often overcome three additional obstacles:

1. Cultural differences/stakeholder bias.
2. Physical distance.
3. Poor communications links.

Stakeholder Perception

The saying "perception is reality" is true in IT projects. How your stakeholders perceive your credibility and competence are the two key ingredients to achieve trust. What's the difference between credibility and competence? Credibility is "earned believability," and is an emotional response. (example: stakeholders believe what you're saying). Competence, on the other hand, is measurable by proving knowledge and capability. (example: executing a task successfully). Our goal is to achieve both credibility and competence.

Five ways to gain credibility

1. Speak convincingly.
2. Have confidence in yourself.
3. Know your material.
4. Be honest and sincere.
5. Listen.

Three ways to achieve competence

1. Educate yourself.
2. Bring in subject matter experts to the table.
3. Deliver as expected.

Communications Links

If you've ever been on the receiving end of line static and signal dropouts during overseas conference calls, you can understand why stakeholders are reluctant to work with offshore resources. I've attended a lot of meetings where important technical points presented by the offshore team were unclear because of a bad telephone line. Sometimes customers are too polite or too embarrassed to ask for repeated items, and they fill in the blanks. In the IT world, if an important point is missed or misinterpreted, it could cause all sorts of problems.

A poor communications link increases the perception of distance and progressively erodes the confidence of the client. A good communications link achieves the opposite. When using clear IP voice links, offshore resources almost sound like they are in the same room.

Four tips for better communications

1. Do a "radio check." Offshore service providers should inspect/test the quality of communications links to validate

clarity regularly. Should link quality be found substandard, take action to improve it.

2. Use an IP based communications link or meeting application (Skype, Zoom, WebEx) when attending customer meetings from remote locations. Sufficient Internet bandwidth is mandatory for good IP communications.

3. Use good quality headsets to improve vocal clarity and eliminate feedback. I find headsets with a single earpiece/boom microphone are best as they allow you to hear your voice clearly when speaking. The boom should be positioned correctly in front of your mouth.

4. Offshore resources should avoid standard phone lines or cell phones as these modes of communications are often have degraded voice quality over extended distances.

Geographic Distance

Offshore resources can be unfamiliar with the client's geography. This lack of familiarity becomes evident when planning logistical tasks. Can you locate the site on a map? What are the time zones? Clients take comfort that you're familiar with their business and site locations. Having to ask the client for basic information, despite being legitimate questions, could create a negative perception.

Three things to do ahead of meeting the client

1. If you are working remotely, find out about your client and review the geography ahead of planning meetings to understand their geographic location.

2. Know the time zone(s) of each worksite, and the time zone(s) of those stakeholders attending meetings.

3. Look up their company on the Internet and try Google maps (street view) to take a look at their facility.

Cultural Differences

Cultural differences are a challenge in situations when delivering services from offshore locations. Things like management styles, communications styles, etiquette, decision, and conflict resolution may vary from culture to culture. North American onshore stakeholders can expect to experience a different dynamic with offshore delivery from India, for example, compared to a team based in Ireland. These variances require understanding by both parties to achieve a good working relationship.

When resources from three or four countries form a project team, things can get even more complicated. In these cases, the project manager must have the skills of a diplomat. One problem is client stakeholders often expect offshore resources to deliver like they are locally based, which isn't a realistic expectation. Vendors can control some aspects of offshore delivery through universal global processes, but cultural differences can't be managed the same way.

How do we then deal with cross-cultural differences? I believe any organization conducting international business should formally train its staff (on both sides of the pond) in each-others cultural ways. This training should detail specific nuances between cultures and cover things like communications, etiquette, and management styles. Knowing what behaviours to expect allows us to identify, understand, and tackle related challenges in a respectful way.

Workload

I've found offshore resources are often overloaded, and their management continues to pile more work on top of existing work. The administration seems to lack visibility or empathy when assigning work. On top of all this, offshore resources have trouble saying "no," and keep accepting more workload to the point of exhaustion.

To do a job effectively, one must have sufficient runway to think and deal with unplanned problems. If tasked with too many projects, then it's impossible to excel on all of them. Something always suffers. This issue is often a two-phase problem. First, the resource in question feels they can't resist taking on more workload out of fear for losing their job, and secondly, their management has their head in the sand and no idea what's going on. You're not doing you or your company any favours when you take on too much work.

Four ways to manage your workload

1. If you manage a portfolio of projects, keep management aware of your workload by giving them regular updates on their status. A project/program portfolio dashboard is a great way to accomplish this.
2. Resist accepting new projects exceeding your workload capability. Point out how the new workload places existing projects and customer satisfaction at risk by stretching you too thin.
3. Block off some time each day or week to get things done. If you have a meeting scheduler observable by others, just set the time aside as "admin," You'll avoid getting invited into meetings during that time slot.
4. When having trouble coping, have a frank conversation with your manager. They may be unaware of your struggles, and they should be willing to work with you to help you get back into a manageable workload situation.

Key Points

☐ Customers prefer local resources, but it's not always realistic.

☐ Offshore resources face more challenges than local resources.

☐ Perceptions are crucial and offshore resources need to focus on demonstrating credibility and competence to remote stakeholders.

☐ The audio quality of offshore links is vital.

☐ Cultural differences are a reality of globalization, which cross-cultural teams need to understand and respectfully accommodate.

☐ Managers must avoid overloading offshore resources, and offshore resources must alert management when excessive workload impacts their ability to do a proper job.

FIVE ESSENTIAL RELATIONSHIP SKILLS

I n the Information Technology field, it seems most training focuses on methodologies and technical knowledge. I agree these elements should be on the list of your skill sets, but in my opinion, having excellent human relations skills is the most valuable asset you can have. Relationship skills are the hardest to master but have the most significant impact on your success.

Developing these skills requires practice and experience. Memorize the basic principles, and practice them until they become natural behaviours. Using these skills will help you build good relationships with people and be an effective leader. These skills are useable inside or outside of work.

This chapter is not a complete summary of everything you need to know about relationships. A guide to human relations can quickly fill a book, so to keep things brief, this chapter covers the top five skills I think are essential.

Make Others Feel Important

A project team is much like an orchestra, with the project leader being its conductor. The conductor isn't a one-man-band. Without individual musicians, a conductor is useless. If the players aren't inspired, receive poor direction, or have lousy sheet music in front of them, the entire performance will be weak. Everyone on your team is essential, and a good leader should make each member of the team feel this way. We all want to feel important and valued that our work is contributing, and our ideas are worthy of consideration. Making other people feel they are genuinely valued and essential is a key leadership trait.

Four ways to make others feel valued

1. Let team members do their job and recognize them as important players.
2. Thank team members for their inputs. A personal call or message of thanks goes a long way. Publically recognize their contributions.
3. Include others in your conversations. Inclusion applies to social and business situations. In a business meeting, ask for inputs from your team. Call team members out by name and ask what they can add to the discussion. In social situations, don't leave others out of your conversation, give them eye contact and get them involved in the conversation. Ask open-ended questions to avoid simple yes/no answers.
4. Address people by name (first or last as appropriate) and use their name periodically in conversation.

Be Interested

People love to talk about themselves and things that interest them. Taking a genuine interest in others will help you develop relationships. At the appropriate time (perhaps waiting for a meeting to start or at a lunch meeting), ask about hobbies, interests, or if anyone has recently travelled to an exciting place. Sometimes, this is a great "ice breaker" You'll be surprised how much people like to talk about themselves and may find common interests.

Practice Empathy

Put yourself in the other person's shoes and see their perspective on things. This skill is efficient during situations of conflict. It doesn't mean you need to abandon your views, but it gives you a balanced viewpoint. When you acknowledge another person's perspective, it can diffuse conflict and open up dialogue.

Help Others Be Successful

Instead of focusing on yourself, direct your efforts to make others look good and contribute to their success. Whether it's a team member, peer, senior manager or a client, helping them succeed is one of the most potent and rewarding contributions you can make.

Have A Positive Attitude

Attitudes are like a virus. They are incredibly contagious and quickly infect your entire team. I've often noticed that project teams have the same demeanour as the leader. Be as positive as you can, and share your enthusiasm with others.

Key Points

- Make others feel important and valued.
- Show interest in others.
- Practice empathy and place yourself in their shoes.
- Help others be successful.
- Have a positive attitude, it's contagious.

STRESS AND MENTAL HEALTH

S tress is a reality of participating in IT projects. In this business, employees have multiple projects, excessive workload, conflicting priorities, unreasonable deadlines, nitpicky managers, wild and crazy processes, unclear scope, difficult people, family problems, health issues and global calamity. The list can go on and on. Burnout is a by-product of stress and is often like an aviation accident, that is, usually caused by several factors when combined, lead to the individuals themselves crashing.

Stress manifests itself mentally, physically or both. Mental stress is often invisible to outside observers like your manager or family members because you can appear to be functioning normally, but instead, you are tearing up on the inside. Stress will negatively impact your life, often leading to anxiety and depression and will eventually affect your family and colleagues' lives. In business, the pressure is costly because it impacts productivity and customer satisfaction. As an individual, you need to be able to identify and act upon stress.

Thirteen signs of stress

1. Feeling tired all the time.

2. Irritability.
3. Loss of enthusiasm for work.
4. Trouble getting out of bed in the morning.
5. Memory loss.
6. Confusion.
7. Nightmares.
8. Can't sleep.
9. Feel depressed.
10. Experience family/spouse relationship issues.
11. Have violent or suicidal thoughts.
12. Experience anxiety.
13. Don't feel like yourself.

Should you have any of the above symptoms for longer than a few days, then see your doctor.

Peaks And Valleys

Realize that its normal to experience peaks and valleys of stress in your work. Most tough times are temporary, and they'll pass. When you hit a rough spot, remind yourself that things will get better, and you'll get through it. A positive attitude will carry you through these brief periods.

Take A Break

Step back from the situation. Take a short break from work when you're stressed; get away from your work environment. Go for a 30-minute walk, and you'll often find you 'll cool down quickly. A break will give you time to think and get a fresh perspective on things.

Get Rest

Rest is a requirement for healthy living. Humans need eight hours of sleep to function correctly, and we also need some time for ourselves to pursue personal interests. Your work hours should be limited, so try and set aside at least four hours of time each day for yourself. We all need time to wind down after a working day.

Speak To Someone

When you feel chronic stress, talk to a friend, partner or colleague about it. Sometimes speaking with a sympathetic soul is all you need to realize you're not alone. Colleagues often have good advice to share to help you through your rough time. Speaking to your manager about your challenges is also recommended. Compared to speaking with a friend or colleague, talking with your manager is difficult because it may feel like an admission of weakness or failure. We're all human. If your job is creating stress in your life, your manager should be there to help you. If your manager isn't there to support you, then it's time to find a new manager or a new company.

Write It Down

Writing down your problems is an excellent way of getting them in front of you. By visualizing your challenges, you can make a better assessment of their magnitude and priority.

Three tips to confront your problems

1. Write each of your challenges on separate post-it-notes, and stick them to the wall in your office. As you solve each problem, remove it from the wall.
2. Create and maintain a list of your challenges.
3. If you're having an extended rough time, keep a diary and document how you feel. This way, you can measure your progress from day-to-day and spot ongoing trends.

Often, by visualizing your challenges, things are not as bad as they appear (or feel). Maintaining a list of open problems and checking off completed items allows you to see progress as you chip away at them, which can be wonderfully therapeutic.

Pet Therapy

It's proven pets can help lower your stress levels. When you are stressed, seek your cat or dog and play with them for a few minutes. Their response will be uplifting, and you will feel better.

It's Okay To Admit You're Challenged

Each one of us has different sets of skills and knowledge. We cannot possibly know everything or be good at everything. It's okay to have challenges beyond your capability, and it's okay to acknowledge it.   When you hit a brick wall, admit you need help and seek it out. I know this is hard, but never be embarrassed to let your employer know you are having problems and need help.

Get A Hobby

Having a hobby or outside interest is a great diversion from work. Whether your hobby is listening or playing a musical instrument, travelling, hiking or biking, anything different and has your mind working in a different direction than "work" is often therapeutic. A surprising number of IT people I know don't have any outside interests. Try and take some time to get some variety in your life. Take up a hobby, take a trip and get away from work whenever possible. Your brain likely needs a rest.

What Managers Must Do For Staff

As a manager, you have a responsibility for the mental and physical welfare of your resources. If you are aware members your staff are operating in a stressful situation as part of their job duties action on your part is required.

Eight ways managers can help employees with stress

1. Call them and find out how they are doing.
2. Show genuine concern and ask how you can help.
3. Put yourself in their place; how would you feel?
4. Listen carefully. You may need to dig deep or read between the lines.
5. Take affirmative action to help them and get executive support if needed.
6. Get creative if necessary.
7. Publically recognize their efforts.
8. Follow up.

Chinese Water Torture

Stress is often aggravated by having to sustain monotonous workloads over time. It's kind of like the sinister "Chinese water torture" method I saw on TV shows when I was a kid. Water is dripped onto the scalp of the unfortunate victim and repeated until the person goes insane eventually. By itself, the drip is painless, but the cumulative effect of the dripping becomes unbearable. To solve this challenge, you must break the monotony.

Four ways to relieve water torture pain

1. Splitting workloads with other project resources
2. Move the impacted person into a new role with a different focus.
3. Inject variety into the work.
4. Change the environment or conditions.

Come Up With An Action Plan

A good strategy for overcoming stress is to come up with an action plan. Make a list of the symptoms and causes of your anxiety and develop an action plan to overcome it.

Personal Health

Good health is one of the best defences against stress. Many jobs in our profession are behind a desk, and it seems we get out less and less these days. More work is being performed remotely from home and as a result, maintaining our health is more important than ever. It's effortless to become complacent and pack on some pounds sitting behind a computer screen all day.

When working from home, the kitchen is often close to your desk, which sometimes makes healthy eating a challenge. Ensure you are getting regular physical checkups and ask your doctor for an assessment of your physical condition. If needed, ask for help to create a plan to get back on track.   A proper diet and exercise are the foundations for good health. There are many resources on this topic, and you might need to dig a little to find an approach that works best for you.

Key Points

☐ Recognize the signs of stress.

☐ Be positive; stress is often temporary.

☐ Take a short break outside the work environment.

☐ Get rest; a healthy lifestyle is vital.

☐ Speak to someone about how you're feeling.

☐ Write it down.

☐ Pets can help relieve stress.

☐ It's okay to admit you're having a challenging time, and you'll likely find you're not alone.

☐ Get a hobby to provide a different focus, pursue a personal passion.

☐ Seek medical help if symptoms persist.

☐ Managers must recognize stress and take action when workplace stress impacts staff well-being.

EPILOGUE

What does the future hold for project managers? In the early 21st century, we might think we are advanced, but we're still in the "wild west" days of project management. Perhaps that's a good thing, or maybe it's a bad thing. Despite having lots of cool project software and tools, they still require manual inputs from someone on the project team. We have applications to help us communicate over distances easier, but they're not perfect. Major stumbling blocks are data security and integration, especially in sensitive collaborative projects.

Future Vision

I believe we'll see the emergence of Artificial Intelligence (AI) in project management and the development of self-serve interfaces for customers. As a first step, manufacturers will begin developing their products from the ground-up to facilitate AI management. AI-based project applications will interface with these new-generation hardware. They will eventually manage most aspects of complex projects, including planning, scheduling, logistics, resource assignment and execution.

It's entirely likely that the project manager of the future will be a virtual entity who will present itself in any image of the stakeholders choosing (an alien, a dog, a cartoon character). Most of the stakeholders themselves may eventually become virtual so that we could have an entirely artificial team with few human participants.

For the foreseeable future, I believe the most successful project managers will be those who can master human relationship skills first, and software skills second. Proper project management still boils down to communications and the human factor; without this, all the world's most excellent software won't help.

While we remain in the "wild-west days, " I wish you the best of luck and hope that the information in this book will help give your project wings!

ACKNOWLEDGEMENT

I wish to thank my good friends Ron Hamilton, Milt Joneson, Jeff Keimel, Sandeep Patil and Gary Williams, for their contributions to this book.

Also, I thank all the colleagues and customers I've had the pleasure of working with through the years, some of whom were mentors and provided me with the knowledge and experience I've shared here.

ABOUT THE AUTHOR

Robert Forsyth

Robert Forsyth (PMP, ITIL) lives in Uxbridge, Canada and has worked for 40 years in the information technology field for Siemens, Nixdorf, Olivetti, Data General, EMC Corporation and Dell Technologies. His experience includes Logistics, Field Services Management, Sales, Project Management and Program Management.

www.ingramcontent.com/pod-product-compliance
Lightning Source LLC
Chambersburg PA
CBHW070316240526
45467CB00045B/450